Melchizedek And the Mystery of Fire

A Treatise in Three Parts

Manly P. Hall

MELCHIZEDEK AND THE MYSTERY OF FIRE

ISBN-10: 0-89314-842-3
ISBN-13: 970-0-89314-842-3

Published by

THE PHILOSOPHICAL RESEARCH SOCIETY
3910 Los Feliz Boulevard
Los Angeles, CA 90027 USA

Telephone 323.663.2167
Fax 323.663.9443
Website www.prs.org
E-mail info@prs.org

Printed in the United States of America

CONTENTS

Page

Introduction 4

Part I

Fire the Universal Deity 9

Part II

Man the Grand Symbol of the Mysteries 23

Part III

The Sacred Fire in the Spine and Brain 43

INTRODUCTION

The elaborate rituals of the ancient Mysteries and the simpler ceremonials of modern religious institutions had a common purpose. Both were designed to preserve, by means of symbolic dramas and processionals, certain secret and holy processes, by the understanding of which man may more intelligently work out his salvation. The pages which follow will be devoted to an interpretation of some of these allegories according to the doctrine of the ancient seers and sages.

Every man has his own world. He dwells in the midst of his little universe as the lord and ruler of the constituent parts of himself. Sometimes he is a wise king, devoting his life to the needs of his subjects, but more often he is a tyrant, imposing many forms of injustice upon his vassals, either through ignorance of their needs or thoughtlessness concerning the ultimate disaster that he is bringing upon himself. Man's body is a living temple and he is a high priest, placed there to keep the House of the Lord in order. The temples of the ancients were patterned after the human form, as a study of the ground plan of either the sanctuary at Karnak or of St. Peter's Church at Rome will prove. *If the places of initiation were copies from the body of man, the rituals which were given in the various chambers and passageways symbolize certain processes taking place in the human body.*

Freemasonry is an excellent example of a doctrine intimating, by means of pageants and dramas, that the regeneration of the human soul is largely a physiologic and biologic problem. For this reason the Craft is divided into two parts, *speculative* and *operative Masonry*. In the lodge room, Masonry is speculative, for the lodge is only a symbol of the human organism. Operative Masonry is a series of mystic activities taking place within the physical and spiritual organisms of those who have assumed its obligations.

The possession of the occult keys to human salvation through the knowledge of self is the goal for which the wise of all ages have labored. It was the hope of possessing these secret formulae that strengthened the candidates who struggled valiantly through the dangers and disappointments of the ancient initiations, sometimes actually giving their lives in the quest of truth. The initiations of the pagan Mysteries were not child's play. The Druid priests consummated their initiatory ritual by sending their candidates out upon the open sea in a small unseaworthy boat. Some never returned from this adventure, for if a squall happened to arise, the boat was immediately capsized.

In Central America at the time when the Mexican Indian Mysteries were in their glory, candidates seeking light were sent into gloomy caverns armed with a sword and were told that if for a second they relaxed their vigilance they would meet a horrible death. For hours the neophytes wandered, beset with strange beasts which seemed even more terrible than they actually were because of the darkness of the caverns. At last, wearied and almost discouraged, the wanderers found themselves on the threshold of a great lighted room cut from the natural rock. As they stood, not knowing which way to turn, there was a whir of wings, a demoniacal cry, and a great figure with the wings of a bat and the body of a man passed swiftly just over the candidates' heads, swinging in its hands a great sword with razorlike edge. This creature was called the Bat God. Its duty was to attempt to decapitate the ones seeking entrance to the Mysteries. If the neophytes were off their guard or too exhausted to defend themselves, they died on the spot, but if they had presence of mind enough to ward off this unexpected blow or jump aside in time, the Bat God vanished and the room was immediately filled with priests who welcomed the new initiates and instructed them in the secret wisdom. The identity of the Bat God has been the basis of many arguments, for while it appears many times in Mexican art and in the illuminated codices, no one knew who or what it actually was. It could fly over the heads of the neophytes and was the size of a man, but it lived in the depth of the earth and was never seen save during the Mysteries, although it occupied an important position in the Mexican Indian Pantheon.

The Mysteries of Mithras were also tests of real courage and perseverance. In the rites the priests, disguised as wild beasts and weird composite animals, attacked the aspirants who were passing through the gloomy caverns in which the initiations were given. Bloodshed was not unusual and many lost their lives striving for the Great Arcanum. When the Emperor Commodus of Rome was initiated into the Mithraic Cultus, being a remarkable swordsman, he defended himself so valiantly that he killed at least one of the priests and wounded several others. In the Sabazian Mysteries a poisonous serpent was placed upon the breast of the candidate, who failed in his initiation if he showed any sign of fear. These incidents from the rituals of the ancients give an inkling of the trials through which seekers after truth were forced to pass in order to reach the sanctuary of wisdom. But when we estimate the wisdom which they received if successful, we realize that it was worth the dangers, for from between the pillars of the gates of Mystery came forth Plato and Aristotle and hundreds of others, bearing true witness to the fact that in their day the *Word* was not lost.

The tortures of initiation and the severe mental and physical tests were intended to serve as a process for eliminating those unfit to be entrusted with the secret powers which the priests understood and communicated to the new initiates at the time of their "raising." Those who hung on crosses for nine hours until they became unconscious, as Apollonius of Tyana, initiated in the Great Pyramid, would never reveal the secret teachings through fear of bodily torture, and such as obeyed the order of Pythagoras that unless they remained silent, speaking to no man for five years, they could not enter his school, were not likely to reveal through thoughtless indiscretion any part of the Mystery which it was forbidden that the foolish should know. Because of the great care used in selecting and testing applicants and the remarkable ability to read human nature displayed by the priests, there never was one who betrayed the more important secrets of the temple. For that reason the *Word* remained lost to all save those who still complied with the requirements of the ancient Mysteries, for the law was, *to such as live the life the doctrine is revealed.*

It is unlawful to reveal to the uninitiated the key links to the chain of the Mysteries. It is permissible, however, without breach of confidence, to explain certain of the lesser secrets, a consideration of which will not only vindicate the integrity of the older hierophants but will also reveal part of the Divine mystery of man's nature. The fact cannot be too strongly emphasized that, regardless of claims to the contrary, the operative Arcanum of the temple has never been revealed to the public. A few candidates who went but a little way along the path and who either became discouraged or were eliminated because of their failure to be honest with themselves, have attempted to expose what they knew, but the inherent weakness which prompted them to betray was recognized by their instructors. Therefore, they were given nothing which could actually supply a link to connect the outer teaching with the wisdom of the sanctuary. The world within man—not the world without, was the concern of the Mysteries of antiquity. Hence we are apt to look upon the priests of old as ignorant when compared with ourselves; but while the modern world is mastering the visible universe and raising a colossal civilization, it is ignorant in the fullest sense of the word concerning the identity of that mysterious lodestone of power in the midst of every living thing, without which no investigation could be conducted and no cities built. Man is never truly wise until he has fathomed the riddle of his own existence, and the temples of initiation are the only repositories of that knowledge—a knowledge which will enable him to unfasten the Gordian knot of his own nature. Still, the great spiritual truths are not so deeply concealed as might be supposed. Most of them are exposed to view at all times, but are not recognized because of their concealment in symbol and allegory. When the human race learns to read the language of symbolism, a great veil will fall from the eyes of men. They shall then know truth and, more than that, they shall realize that from the beginning truth has been in the world unrecognized, save by a small but gradually increasing number appointed by the Lords of the Dawn as ministers to the needs of human creatures struggling to regain their consciousness of divinity.

The supreme Arcanum of the ancients was the key to the nature and power of fire. From the day when the hierarchies first descended upon the sacred island of the polar ice cap, it has been decreed that fire should be the supreme symbol of that mysterious, abstract divinity which moves in God, man, and Nature. The sun was looked upon as a great fire burning in the midst of the universe. In the burning orb of the sun dwelt the mysterious spirits controlling fire, and in honor of this great light, fires burned upon the altars of countless nations. The fire of Zeus burned upon the Palatine Hill, the fire of Vesta upon the altar of the home, and the fire of aspiration upon the altar of the soul.

PART I

FIRE THE UNIVERSAL DEITY

Since the earliest times man has venerated the element of fire above all others. Even the most untutored savage seems to recognize in the flame something closely resembling the volatile fire within his soul. The mysterious, vibrant, radiant energy of fire was beyond his ability to analyze, yet he felt its power. The fact that during thunderstorms fire descended in mighty bolts from heaven, felling trees and otherwise dealing destruction, caused the primogenial human being to recognize in its fury the anger of the gods. Later, when man personified the elements and created the multitudinous pantheons which now exist, he placed in the hand of his supreme Deity the torch, the thunderbolt, or the flaming sword, and upon his head a crown, its gilded points symbolizing the flaming rays of the sun. Mystics have traced sun worship back to early Lemuria and fire worship to the origin of the human race. In fact, the element of fire controls to a certain degree both the plant and animal kingdoms, and is the only element which can subjugate the metals. Either consciously or instinctively, every living thing honors the orb of day. The sunflower always faces the solar disc. The Atlanteans were sun worshippers, while the American Indians (remnants of the earlier Atlantean people) still regard the sun as the proxy of the Supreme Light-Giver. Many early peoples believed that the sun was reflector rather than source of light, as is evidenced by the fact that they often pictured the Sun-God as carrying on his arm a highly polished shield, on which was chased the solar face. This shield, catching the light of the Infinite One, reflected (like Lucifer) was chained, but the Greek hero was placed upon the brow of Mount Caucasus, there to remain with a vulture gnawing at his liver until a human being should master the sacred fire and become perfect. This prophecy was fulfilled by Hercules, who climbed Mount Causcasus, broke the fetters of Prometheus, and liberated the friend of man who had been in torture for so many ages. Hercules represents the initiate, who, as his name implies, partakes of the glory of light. Prometheus is the vehicle of solar energy. The divine fire which he brought to men is a mystic essence in their own natures, which they must redeem and regenerate if they would liberate their own crucified souls from the rock of their base physical natures.

According to occult philosophy, the sun in reality is a threefold orb, two parts of its nature being invisible. The globe which we see is merely the lowest phase of the solar nature and is the body of the Demiurgus, or, as the Jews call him, *Jehovah*, and the Brahmins, *Shiva.* The sun being symbolized by an equilateral triangle, the three powers of the solar disc are said to be coequal. The three phases of the sun are called Will, Wisdom, and Action. Will is related to the principle of life, Wisdom to the principle of light, and Action, or Friction, to the principle of heat. By Will the heavens were created and the eternal life continued in supreme existence; by Action, friction, and striving, the earth was formed, and the physical universe molded by the Lords of the Fire Mist passed gradually from its molten condition into its present and more orderly state.

Thus Heaven and Nature were formed, but between these two was a great void, for God did not comprehend Nature and nature did not comprehend the Deity. The lack of intercourse between these two spheres of consciousness was similar to the condition of paralysis in which the consciousness realizes the condition of its body, but owing to the lack of nerve connection it is incapable of governing or directing the activities of the it to all parts of the universe. During the year the sun passes through the twelve houses of the heavens, where, like Hercules, it performs twelve labors. The annual death and resurrection of the sun has been a favorite theme among unnumbered religions. The names of nearly all the great Gods and Saviors have been associated with either the element of fire, the solar light, or its correlate, the mystic and spiritual light invisible. Jupiter, Apollo, Hermes, Mithras, Bacchus, Dionysius, Odin, Buddha, Krishna, Zoroaster (Zarathustra), Fo-Hi, Iao, Vishnu, Shiva, Agni, Balder, Hiram Abiff, Moses, Samson, Jason, Vulcan, Uranus, Allah, Osiris, Ra, Bel, Baal Nebo, Serapis, and King Solomon are some of the numerous deities and supermen whose symbolic attributes are derived from the manifestations of the solar power and whose names indicate their relationship to light and fire.

According to the Greek Mysteries, the gods, gazing down from Mount Olympus, repented that they had made man, and never having given to the primitive creature an immortal spirit, they decided that no harm would be done if the quarreling, dissenting human ingrates were destroyed forever and the place where they had been left vacant for a nobler race. Discovering the plans of the gods, Prometheus, in whose heart was a great love for struggling humanity, determined to bring to mankind the divine fire, which would make the human race immortal so that not even the gods could destroy it. So Prometheus flew to the home of the Sun-God and, lighting a tiny reed with the solar fire, he carried it to the children of the earth, warning them that the fire should always be used for the glorification of the gods and the unselfish service of each other. But men were thoughtless and unkind. They took the divine fire brought them by Prometheus and used it to destroy one another. They burned the homes of their enemies and with the aid of heat they tempered steel, making swords and armor. They grew more selfish and more arrogant, defying the gods, but they could not be destroyed, for they possessed the sacred fire. For his disobedience, Prometheus body. Therefore, between life and action there came a mediator, which was called Light, or Intelligence. Light partakes of both life and action: it is the sphere of blending. Intelligence stood between heaven and earth, for through its medium man learned of the existence of his God and God began His ministration to the needs of men. While both life and action were simple substances, light was a compound, for the invisible part of light was of the nature of heaven and the visible part of the nature of earth. Down through the ages this light is said to have taken upon itself bodies. Although these bodies have borne witness to that light, the great spiritual truth behind the symbol of the embodied light is that in the soul of every creature within whose mind intelligence is born there dwells a spirit which assumes the nature of this intelligence. Every truly intelligent man and woman who is working to spread light in the world is *christ*-ened, or *Light*-ened, by the actual labor which he or she is seeking to perform. The fact that light (intelligence) partakes of the natures of both God and the earth is proved by the names given to the personifications of this light for at one time they are called the "Sons of Men' and at another time the "Sons of God."

The initiate in the Mysteries was always instructed concerning the existence of three suns, the first of which—the vehicle for God the Father—enlightened and warmed his spirit; the second—the vehicle of God the Son—unfolded and broadened his mind; the third—the vehicle of God the Holy Spirit—nourished and strengthened his body. Light is not only a physical element, it is also a mental and spiritual element and in the temple the disciple is told to revere the invisible sun even more than the visible one; for every visible thing is only an effect of the invisible or causal, and as God is the Cause of All Causes He dwells in the invisible World of Causation. Apuleius, when initiated into the Mysteries, beheld the sun shining at midnight, for the chambers of the temple were brilliantly illuminated, although there were no lamps of any kind. The invisible sun is not limited by walls nor even the surface of the earth itself, for its rays being of higher vibratory rate than physical substance, its light passes unimpeded through all the planes of physical substance. To those capable of seeing the light of these spiritual orbs, there is no darkness, for they dwell in the presence of limitless light and at midnight see the sun shining under their feet.

By means of one of the lost arts of antiquity the priests of the temple were able to manufacture lamps which would burn for centuries without replenishment. The lamps resembled what is commonly called the "virgin lamps," or those carried by the Vestal Virgins. They were a little smaller than a human hand and, according to available records, their wicks were made of asbestos. It has been maintained that these lamps have burned for a thousand years or more. One of them was found in the tomb of Christian Rosencreutz which had burned for 120 years without the supply of fuel being diminished. It is supposed that these lamps (which, incidentally, burned in hermetically sealed vaults without the aid of oxygen) were so constructed that the heat of the flame extracted from the atmosphere some substance which took the place of the original fuel as rapidly as the mysterious oil was consumed.

Hargrave Jennings has collected numerous references concerning the times when and places where these lamps have been found. In the majority of cases, however, they went out shortly after the vaults were opened or else were broken in some strange way, so the secret was not discovered. Concerning these lamps, Mr. Jennings writes: " The ancient Romans are said to have preserved lights in their sepulchers many ages by the *oiliness of gold* (here steps in the art of the Rosicrucians), resolved by hermetic methods into a liquid substance; and it is reported that at the dissolution of monasteries, in the time of Henry the Eighth, there was found a lamp which had then burned in a tomb from about three hundred years after Christ—nearly twelve hundred years. Two of these subterranean lamps are to be seen in the Museum of Rarities at Leyden, in Holland. One of these lamps, in the Papacy of Paul the Third, was found in the Tomb of Tullia (so named), Cicero's daughter, which had been shut up fifteen hundred and fifty years."

Madame Blavatsky, in Isis Unveiled, gives a number of formulae for the making of ever-burning lamps, and states in a footnote that she herself saw one, made by a disciple of the hermetic arts, which had burned steadily without fuel for six years previous to the publication of her book.

The ever-burning lamp was of course a most appropriate symbol of the Eternal Fire in the Universe, and while chemistry has denied the possibility of manufacturing them, the fact that many have been made and seen over a period of thousands of years is a warning against dogmatizing. In Tibet, the Lama magicians have discovered a system of lighting rooms by means of a luminous ball of phosphorescent, greenish-white color, which increases in luminosity when ordered to do so by the priests, and after the departure of those who are in the chamber it gradually becomes fainter until only a spark remains, which burns continuously.

This apparent miracle is no more difficult to explain than another performed by the Tibetans. There is in Tibet a sacred tree which sheds its bark annually, and as the old bark peels off an inscription written in Tibetan characters is found upon the new bark underneath. These secrets of so-called savage and primitive peoples incessantly refute the ridicule with which Caucasians almost invariably view the culture of other races.

The Druid priests in Britain, recognizing the sun as the proxy of the Supreme Deity, used a ray of solar light to start their altar fires. They did this by concentrating the ray upon a specially cut crystal or aquamarine, set in the form of a magic brooch or buckle upon the front of the belt of the Arch-Druid.

This brooch was called the "Liath Meisicith" and was supposed to possess the power of drawing the divine fire of the gods down from heaven and concentrating its energies for the service of men. The buckle was, of course, a burning glass. Many of the nations of antiquity so revered the fire and light of the sun that they would not permit their altars to be lighted by any other means than the concentration of the sun's rays through a burning-glass. In certain of the ancient temples, specially arranged lenses were placed in the ceiling at various angles so that each year at the vernal equinox the sun at high noon would send its rays through these glasses and light the altar fires which had been specially prepared for this occasion. The priests considered this process equivalent to the gods having actually lighted the fires themselves. In honor of Hu, the supreme Deity of the Druids, the people of Britain and Gaul celebrated an annual lighting of fires on what they termed Midsummer's Day.

One of the reasons why the mistletoe was sacred to the Druids was because many of the priests believed that this peculiar parasitic plant fell to the earth in the form of lightening bolts and wherever a tree was struck by lightning the seed of the mistletoe was placed within its bark. The great length of time the mistletoe remained alive after being cut from the tree had much to do with the veneration showered upon it by the Druids. That this plant was also a powerful medium for the collection of the mysterious cosmic fire circulating through the ethers was discovered by the early priests, who valued the mistletoe because of its close connection with the mysterious astral light, which is in reality the astral body of the earth. Concerning this, Eliphas Levi writes in his *History of Magic*: "The Druids were priests and physicians, curing by magnetism and charging amulets with their fluidic influence. Their universal remedies were mistletoe and serpents' eggs, because these substances attract the astral light in a special manner. The solemnity with which mistletoe was cut down drew upon this plant the popular confidence and rendered it powerfully magnetic.** The progress of magnetism will some day reveal to us the absorbing properties of mistletoe. We shall then understand the secret of those spongy growths which draw the unused virtues of plants and become surcharged with tinctures and savors. Mushrooms, truffles, gall on trees and the different kinds of mistletoe will be employed with understanding by a medical science, which will be new because it is old."

Certain plants, minerals, and animals have been held sacred among all nations of the earth because of their peculiar sensitiveness to the astral fire. The cat, sacred to the city of Bubastis in Egypt, is an example of a peculiarly magnetized animal. Anyone stroking the fur of a domestic cat in a dark room can see the electrical emanations in the form of green phosphorescent light. In the temples of Bast, sacred to the cat goddess, three-colored cats were viewed with unusual veneration, as was any member of the feline family whose two eyes were of different colors. Lodestone and radium in the mineral kingdom and various parasitic growths in the plant kingdom are strangely susceptible to the cosmic fire. The magicians of the Middle Ages surrounded themselves with certain animals, such as bats, cats, snakes, and monkeys, because they were able to borrow the power of the astral light from these creatures and appropriate it to their own uses. For this same reason the Egyptians and certain of the Greeks kept cats in the temples and serpents were always in evidence at the oracle of Delphi. The auric body of a snake is one of the most remarkable sights that the clairvoyant will ever see, and the secrets concealed within its aura demonstrate why the serpent is the symbol of wisdom among so many nations.

That Christianity has preserved (in part at least) the primitive fire worship of antiquity is evident in many of its symbols and rituals. The incense burner so often used in Christian churches is a pagan symbol relating to the regeneration of the human soul. The incense burner represents the body of man. The incense within the burner, made from the extracted essences of various plants, represents the life forces within the body of man. The flaming spark burning in the midst of the incense is emblematic of the spiritual germ concealed in the midst of the material organism of man. This spiritual spark is an infinitesimal part of the divine flame, the Great Fire of the Universe, from Whose flaming heart the altar fires of all His creatures have been lighted. As the spark of life gradually consumes the incense, so the spiritual nature of man through the process of regeneration gradually consumes all the gross elements of the body, transmuting them into soul power —symbolized by the smoke. Although smoke is actually a dense and physical substance yet light enough to rise in clouds, so the soul is actually a physical element, but through purification and the fire of aspiration it has taken upon itself the nature of intangible atmosphere; though composed of the substance of earth, it becomes light enough to rise as a fragrant odor into the presence of Deity.

While some authorities have held that the form of the cross was derived from the ancient Egyptian instrument called the "Nilometer" and used for measuring the inundations of the Nile, others hold the opinion that the symbol had its origin in the two crossed sticks used by primitive peoples to generate fire by friction. The use of the bell towers and campaniles in the construction of the cathedrals of medieval Christianity, also the more familiar conventionalized church steeple, may be traced to the fire obelisks of Egypt, which were placed in front of the temples sacred to the superior deities. All pyramids are symbols of fire, while the heart used on valentines is merely and inverted candle flame. The Maypole had its origin in similar antiquity, where it is both a phallic symbol and an emblem of cosmic fire.

The prevailing custom of having churches face the East is, of course, further evidence of the survival of sun worship. Practically the only branch of the human race that does not observe this rule is the Arabic. The Mohammedans face their mosques toward Mecca, but still have their appointed hours of prayer governed by the sun. The rose windows and ivy covered walls are survivals of pagandom, for ivy was sacred to Bacchus because of the shape of its leaf, and this plant was always allowed to trail over the walls of the temple sacred to the Greek solar deity. The golden ornaments upon the altars of Christian churches should remind the philosophical observer that gold is the sacred metal of the sun, because (according to alchemists) the sun ray itself crystallized in the earth, thus forming this precious metal —which, incidentally, is still being made. The candles so often seen adorning the altars, and most frequently appearing in an uneven number, are a reminder that the uneven numbers are sacred to the sun. When three candles are used, they symbolize the three aspects of the sun: sunrise, noon, and sunset, and are thus emblematic of the Trinity. When seven are used, they represent the planetary angels called by the Jews "Elohim", whose numerical and Qabbalistic values are also seven. When the even numbers 12 or 24 appear, they represent the signs of the zodiac and the spirits of the hour of the day, called by the Persians the "Izeds." When only one light is shown, it is the emblem of the Supreme Invisible father, Who is One, and the little red lamp ever burning over the altar is an offering to the Demiurgus—Jehovah, or the Lord Builder of Forms.

What oil is to the flame, blood is to the spirit of man. Therefore, oil is often used in anointing, for it is a fluid sacred to the solar power. Because oil contains the life of the sun, it is used in large quantities in far Northern lands where it is necessary to generate an abundance of body heat. Hence, the proclivity of the Eskimos for eating tallow candles and whale oil.

The actual word "Christ" is itself sufficient proof that fire and the worship of fire are the two most essential elements of the Christian faith. The rays of light pouring from the sun were viewed by the ancients as the blood of the Celestial Lamb which at the vernal equinox died for the sin of the world and redeemed all humanity through its blood (rays).

The Mystery Schools of ancient Egypt taught that the blood was the vehicle of the consciousness. The spirit of man traveled through the bloodstream and therefore was not actually located in any one part of the compound organism. It moved through the body with the rapidity of thought, so that consciousness of self, cognition of externals, and sense perception could be localized in any part of the body by the exercise of the will power. The initiates view the blood as a mysterious liquid, somewhat gaseous in nature, which served as a medium for manifesting the fire of man's spiritual nature. This fire, coursing through the system, animated and vitalized all parts of the form, thus keeping the spiritual nature in touch with all of its physical extremities. The mystics looked upon the liver as the source of the heat and power in the blood. Hence it is significant that the spear of the centurion should pierce the liver of Christ and the vulture should be placed over the liver of Prometheus to torment him throughout the ages.

Occultism teaches that it is the presence of the liver which distinguishes the animal from the plant and that mystically certain small creatures having power of motion but no liver are actually plants in spiritual consciousness. The liver is under the control of the Planet Mars, which is the dynamo of this solar system and which sends a red animating ray to all the evolving creatures within this solar scheme. The philosophers taught that the planet Mars, under the control of its regent Samael, was the transmuted "Sin-Body" of the Solar Logos which originally had been the "Dweller on the Threshold" of the Divine Creature whose energies are now distributed through the fire of the sun. Samael, incidentally, was the fiery father of Cain, through whom a part of humanity has received the flame of aspiration and are thus separate from the sons of Seth, whose father was Jehovah.

The Egyptians considered the juice of the grape to be more nearly like human blood than any other substance. In fact, they believed that the grape secured its life from the blood of the dead who had been buried in the earth. Concerning this subject, Plutarch writes as follows: "The priests of the Sun and Heliopolis never carry any wine into their temples, * * * and if they made use of it at any time in their Libations to the Gods, it was not because they looked upon it, as in its own nature acceptable to them; but they poured it upon their altars as the blood of those enemies who formerly had fought against them. For they look upon the vine to have first sprung out of the earth after it was fattened with the carcasses of those who fell in the wars against the Gods. And this, say they, is the reason why drinking its juice in great quantities makes men mad and beside themselves, filling them as it were with the blood of their own ancestors—" (*Isis and Osiris.*) The magicians of the Middle Ages were aware of the fact that they, by their occult powers, could control any person by first securing a small amount of his blood. If a glass of water be left over night in a room where someone is sleeping, the next morning the water will be impregnated to such an extent with the psychic radiations of that person that anyone understanding the modus operandi may find contained in the water a complete record of the life and character of the one who occupied the room. These records are transmitted and preserved in a subtile substance which the medieval transcendentalists called the *astral light*, an ever-present all-pervading fiery essence, which preserves intact the record of everything transpiring in any part of Nature.

The streaming rays pouring from the face of the sun have caused it to be associated with the lion, because of the shaggy mane of this king of beasts. The golden-haired Savior Gods of many nations subtly signify by their uncut locks the solar radiations. The sun was the king of heaven, and earthly rulers desiring to advertise their terrestrial power delighted to be considered as "Little Suns," their vassals being viewed as planets basking in the glory of the central light. The highest of each kingdom in Nature was also considered symbolic of the sun. Hence the scarab beetle, being the most intelligent of all insects, the eagle the most aspiring of all birds, and the lion the strongest of all beasts, were considered fitly symbolic of the solar disc. Thus the Moguls chose the lion for their standard, while Caesar and Napoleon used the eagle to symbolize their dignity. The crowns of kings were originally bands of gold with radiating points to symbolize that they partook in part of the divine power vested in the sun. As time went on, the crown was conventionalized. Its surface was encrusted with jewels, a number of its points were changed, and its evident resemblance to the sun was lost.

The halo so often seen pictured around the heads of both Christian and pagan deities and saints is also emblematic of the sun power. According to the Mysteries, there comes a time in the spiritual unfoldment of man when the mysterious oil which has been moving slowly up the spinal column finally enters the third ventricle of the brain, where it becomes beautifully golden in color and radiates in all directions. This radiance is so great that it can-not be limited by the skull and it pours out from the head, especially from the back of the neck where the uppermost vertebra of the spine articulates with the condyles of the occipital bone. It is this light pouring out in a fan-shaped aura around the posterior part of the head that has given rise to the halos of saints and the nimbus so often used in religious art. This light signifies human regeneration and it forms part of the auric bodies of man.

These auras have greatly influenced the color and form of the garments used in religious ceremonials. The robe of blue and gold which Albert Pike speaks of and the vestments of the different degrees in the hierarchies of all religious orders are symbolic of these invisible emanation forms which surround man, their colors changing with his every thought and feeling. By means of these auras the priests and philosophers of the ancient world chose those disciples who would do credit to their teachings. The "Robes of Glory" of the High Priest of Israel are all symbolic, as Josephus with his Oriental instruction has shrewdly noted. The plain white linen symbolizes the purified physical nature; the many-colored garments represent the astral body, the blue raiment the spiritual nature, and the violet the mind, for it is a color made up of two shades, one spiritual and the other material.

In the Egyptian Mysteries it was not uncommon to show the rays of the sun ending in human hands. One of the chairs recently found in the tomb of Tutankhamen showed a sun with its rays ending in human hands. Among the ancients the hand was the symbol of wisdom, because it was used to raise the fallen, and no man is so low in his estate as an ignorant man. The physical proclivities of the sun and its water-drawing power were used to symbolize a spiritual process in which the divine nature of man was raised, or illuminated, and drawn upward by the heat of the sun, these emanating rays spreading the threefold spiritual power as love, wisdom, and truth.

PART II

MAN THE GRAND SYMBOL OF THE MYSTERIES

Pythagoras said that the universal Creator had formed two things in His own image: The first was the cosmic system with its myriads of suns, moons, and planets; the second was man, in whose nature the entire universe existed in miniature. Long before the introduction of idolatry into religion, the early priests to facilitate their study of the natural sciences caused the statue of man to be placed in the sanctuary of their temples, using the human figure to symbolize the Divine Power in all its intricate manifestations, Thus the priests of antiquity accepted man as their text-books, and through the study of him learned to understand the greater and more abstruse mysteries of the celestial scheme of which they were a part. It is not improbable that this mysterious figure standing over the primitive altars was made in the nature of a mannequin and, like certain emblematic hands in the Mystery Schools, was covered with hieroglyphs either carved upon its surface or painted there-on with everlasting pigments. The statue may have opened, thus showing the relative positions of the organs, bones, muscles, nerves and other parts.

The present generation is prone to underestimate the knowledge of anatomy possessed by ancient races. Owing to destruction by time and vandalism, the available records do not adequately represent the learning of antiquity. Professor James H. Breasted, archeologist of the University of Chicago, recently stated that his investigations showed that the learned doctors of Egypt during the 18th dynasty—that is, about seventeen centuries before Christ—had a medical knowledge comparable to that of the twentieth century. Professor Breasted is quoted as follows: "For instance in it (the Edwin Smith papyrus, an early scientific document) the word 'brain' appears for the first time recorded in human speech, and there is evidence that the Egyptians understood the localization of brain control of muscles—a knowledge that has only been rediscovered within the last generation."

The knowledge which the Egyptian priest-physicians possessed concerning the functions of the several parts of the human body not only equaled that of many modern scientists, but as regards those functions and powers concerned with the spiritual nature of man and organs and centers through which the spiritual essences control the body, their knowledge exceeded that of the modern world.

During ages of research, much was contributed to fundamental principles of the early philosophers, and at the time Egypt reached the crowning glory of her civilization the manikin was a mass of intricate hieroglyphs and symbolic figures. Every part had its secret meaning. The measurements of this stone figure formed a basic standard by means of which it was possible to measure all parts of cosmos. It was a glorious composite emblem of all the knowledge possessed by the sages and hierophants of Isis, Osiris, and Serapis.

Then came the time of idolatry. The Mysteries decayed from within. The secret meanings were lost and none knew the identity of the mysterious man who stood over the altar. It was only remembered that the figure was a sacred and glorious symbol of the universal power. This figure came to be looked upon as a god, the one in whose image man was made. The secret knowledge of the purpose for which the manikin was constructed being lost, the priests worshipped the actual wood and stone, until finally their lack of spiritual understanding brought the temple down in ruins about their heads and the statue crumbled with the civilization which had forgotten its meaning.

Today the great faith of the white race—Christianity—is served by a great number of honest, sincere, devout men and women. While devoted to their task, they are only partly efficient, because the majority of them are totally ignorant of the fact that so-called Biblical Christianity is an allegory concerning the true spirit of Christianity and of that esoteric doctrine evolved in the temple by the initiated minds of pagandom, and promulgated to serve the religious needs of the human race. Today this faith is served by millions and understood by only a handful, for while the Mystery temple no longer exists as an institution on the corners of the streets as it did in the ancient world, the Mystery School still exists as an invisible, philosophical structure. It admits into the knowledge of its secrets only a few, permitting the great mass to enter only the outer courtyard and make its offering upon the brazen altar. Christianity is essentially a Mystery School, but most of its adherents do not understand it well enough to realize that there are secrets concealed behind the parables and allegories which are an important part of its dogma.

Why should Christianity not be a Mystery School? Its founder was an initiate of the Essenian Mysteries. The Essenes were disciples of the great Pythagoras and were also connected with the Secret Schools of India. The Master Jesus was himself a hierophant, deeply versed in the ancient Arcanum. St. John by his writing proves himself to be acquainted with the ritualism of the Egyptian cult, and it is contended that St. Matthew was the teacher of Basilides, the immortal Egyptian sage, co-founder with Simon Magus of Gnosticism, the most elaborate system of Christian mysticism that has ever evolved from the main stem of St. Peter's church. During its early history in Rome,

Christianity was in constant contact with Mithraism, the fire philosophy of Persia, from which it borrowed no small part of its rituals and ceremonials.

If Christianity were looked upon less as a church and more as a Mystery School, the modern world would rapidly gain a clearer understanding of its tenets. Every priest of Christendom, every minister of the gospel should be an anatomist and a physiologist, a biologist and a chemist, a physician and an astronomer, a mathematician and a musician, and above all, a philosopher. By a philosopher we mean one who could study intelligently all these different lines of thought and discover the interrelationship existing between them, and use all the arts and sciences as methods by which to interpret the magnificent emblematic pageant and mystery drama of the Christian faith. If they were to intelligently consider the secrets handed down from the priests of pagan antiquity (whose stupendous genius soared far above the rutted prejudices of modern thought), they would make a number of important discoveries.

First of all, they would discover that in the present translation of both the Old and New Testaments are numerous mistakes, owing to the fact that the translators were not spiritually competent to interpret the secret mysteries of the Hebrew and Greek languages. They would find numberless contradictions caused by misunderstanding and would also discover that the so-called Apocryphal books (rejected as uninspired) contain some of the most important keys which have descended to us from antiquity.

They would learn that the Old Testament was not to be considered literally; that concealed between its lines were certain secret teachings without which the true meaning of the Hebrew writings cannot be discovered. They would no longer laugh at the pagans for their plurality of gods, for they would discover that they themselves, if faithful followers of their Scriptures, are polytheists. The word "Elohim" as used in the early chapters of Genesis and translated "God," is a masculine-feminine plural word meaning a number of gods who are androgynous, and not one Supreme Deity. They would realize that Adam was not a man, but a species—a race of creatures; they would also realize that the Garden of Eden was not located in Asia Minor.

Even if some men knew these things to be true, a great part of humanity would still reject them, because they would disagree with the accepted traditions, venerated not because they are true but because they have been accepted for generations. They would crown their discoveries by a realization that the Holy Land of all nations is the human body; that this is sacred earth, consecrated to the gods. They would realize that their own bodies were the Holy Sepulchres that have long been in the hands of the infidel, and they would realize that there is no infidel of any race half so heartless as the infidel which dwells in the heart of man himself; that there is no enemy to the faith like the lower nature of the individual; there is no Judas like selfishness, no betrayer like ignorance, no tyrant like pride, no Red Sea to be crossed like that which comprises the emotional nature of man, surging outward form the red blood-forming centers in the human liver.

If the modern theologians could see the ancient mannequin over the altar, they would clearly understand all this, but not realizing that there is a secret doctrine they do not seek for it. Yet who can read the Book of Ezekiel and Revelation and not realize that the Beloved Disciple, John, transcending all the others in his vision, was indeed lifted up, or "raised" as the modern Mason might say, and beheld the pageantry of the Mysteries. The allegories of St. John are drawn from every religion of the ancient world. The drama which he unfolds in Revelation is synthetic and therefore truly Christian in that it includes the great teachings of all ages. Some believe that God has not willed that man should understand the mystery of his own destiny, but let these recall those immortal words; "There is nothing concealed that shall not be revealed; there is nothing hidden that shall not be made known." This being true, let us take up the labor of solving, or unveiling, of reconstructing. Following in the footsteps of the illumined of all ages, we too shall discover truth, by following the winding stairs up which the candidates of every nation and religion have passed, wearing ruts in the stones.

The spirit of man is a tiny ring of colorless fire from which pour streamers and rays of scintillating force. By a mystic process the rays build bodies around that central formless germ, and man dwells in the midst of these bodies, controlling them by waves of force in a manner difficult to appreciate unless one is familiar with the occult constitution of man.

This ring of invisible flame is the eternal fire, the spark from the Infinite Wheel, the birthless, deathless, eternal center which includes within itself all that it has ever been, all that it is, and all that it ever shall be. This germ dwells in the state Eternity, for to this immortal spark time is illusionary, distance is nonexistent, joy and sorrow are unknown, for concerning its function and consciousness all that can be said is that "It is." While other things come and go, *It is.*

This germ of immortality enters into the embryo at the time of quickening and passes out at the moment of death. With its coming, heat is generated; with its leaving, heat is withdrawn. As the flaming orb of the sun is in the midst of the solar system, so this flaming ring of spirit is in the midst of the bodies of man. It is the altar fire which never goes out, and to the service of this divine flame the wise of all nations have consecrated themselves, for in this flame lies all perfection and the possibility of ultimate attainment. This flame manifests individualities and personalities, but the extracted essences of experience, intelligence, and activity stored up in the individualities and personalities are finally absorbed into this flame, furnishing it with fuel with which it gleams and burns more brightly. From this one altar fire all of the fires in the human body are lighted, like the countless flames which have been started from the sacred fires of the Parsees.

Compare the flaming spirit of man to the light of a candle. First, in the midst of the candle, close to the wick, is a glow nearly colorless. Around this is a ring of golden light, and still further out surrounding the yellow is a deeper orange or reddish flame, which gives off more or less smoke. These three lights—blue, yellow, red—are closely related to the flame in man, for there is a blue, fuel-less light, and there is a yellow light supplied by a pure oil that burns with a steady glow, giving no smoke. Then there is a red flame supplied with a coarser fuel. This is called the consuming fire of the ancients, for in the human body the blue flame is the clearly burning light of reason, illuminating the mind and lighting the darkness of the night, while the red flame is the false light, the fire of passion and lust. It is smoky like the battlefield where hates and fears go up together in one seething, lurid sheet of brick-red flame.

These are the three fires—the fire of divinity, the fire of humanity, the fire of the demons. These three are enshrined within the nature of man, whence their radiance goes forth as the sacred trisyllabic word by which the heavens are created, the earth formed, and the works of evil destroyed. The disciples of the ancient wisdom realized that during the dawn of this earth scheme certain instructions were deposited in safe places by the Sons of the Dawn, or as we call them, the gods, and that after having insured that these doctrines would be preserved for the ultimate salvation of the race, the gods entered into the constitution of man and lost their identity. For this reason it is said that the kingdom of heaven is within you, for the kingdom of heaven includes the divine Father, His Trinity, His seraphim, cherubim, powers, dominations, principalities, thrones, angels, and archangels.

Each of these celestial creatures has contributed something to the nature of man. Through the power of one he feels; through the power of another he sees; through the power of a third he speaks; through the power of a fourth he understands; through the power of the Divine Father he is immortal; through the power of the Trinity he is threefold in his constitution—spiritual, intellectual, physical; through the power of the seraphim the great fires were given to him, while from the cherubim he secured his composite form. Hence these spirits are confined within his own nature until man builds that nature to the point where he releases these cosmic powers through giving them adequate expression and no longer limiting them by his own ignorance and perversion.

In truth, the kingdom of heaven is within man far more completely than he realizes; and as heaven is in his own nature, so earth and hell are also in his constitution, for the superior worlds circumscribe and include the inferior, and earth and hell are included within the nature of heaven. As Pythagoras would say; "The superior and inferior worlds are included within the area of the Supreme Sphere." So all the kingdoms of earthly nature, the minerals, the plants, the animals, and his own human spirit are included within his physical body, and he himself is the appointed guardian spirit of the mineral kingdom and he is responsible to the creative hierarchs for the destiny of the stones and metals.

The infernal world is also part of himself, for within his nature is Lucifer, the Beast of Babylon, Mammon, Beelzebub, and all the other infernal Furies. At the basis of his spine burns an infernal fire, and the Witch's Sabbath so glowingly described by Eliphas Levi can be traced to its source in the lower emotional centers of the human body.

Thus man is heaven, earth, and hell in one, and his salvation is a much more personal problem than he realizes. Realizing that the human body is a mass of psychic centers and that during life the form is crisscrossed with endless currents of energy, that all through the form are sunbursts of electric force and magnetic power, man can be seen by those who know how to see as a solar system of stars and planets, suns and moons, with comets in irregular orbits circling through them. As the Milky Way is supposed to be a gigantic cosmic embryo, so man is himself a galaxy of stars, each of which some day will be a constellation in itself.

Whichever way we look, we find life. Wherever we find life, we find light, for in the midst of all these living things are tiny sparks of immortal splendor. Those whose eyes are chained by earthly limitations see the forms, but to those transcending materiality each life appears as a gleam of immortal splendor. Even the atmosphere is alive with lights, and the clairvoyant passes through spheres of flame. There are lights of a thousand colors and rainbow hues for surpassing in brilliancy the luminosity of the sun, lights a thousand times more varied than the spectrum that we know, color undreamed of, lights so brilliant they cannot be seen but are felt as ringing sounds in the head, lights that must be heard, others like solid columns of fire that must be felt. Wherever the seer gazes he beholds fire. It pours from the stone; it flashes in geometric stars from the petals of flowers, and shoots in waves from the fur of animals. It surrounds man with an aureole of radiance and the earth with a halo of rainbow bands extending miles from its surface. Fire pours light upward through the surface of the earth; it shoots light downward from the empty air; it radiates light outward from the center of everything, and inward from the circumference of everything.

Is it strange that this universal, living splendor was revered? It is man's most perfect symbol of God, for this light is the primary manifestation of the Unmanifested and Eternal One. This eternal fire, burning fuelless in the soul of everything, has been since the beginning of time the most sacred symbol in all the world, for while figures of wood and stone, paintings on canvas, and even songs are more or less expressions of the form, the physical side of Nature, this radiant light, this flaming splendor is symbolic of the spirit, the life, the immortal germ in the midst of form. It was sacred to the Superior Deity and all worshipped it and made offering to it. It was the source, and men worshipped the source, seeking by secret culture handed down through the ages and based upon the instructions of the gods themselves, to make that light shine out more gloriously from within themselves. This is the source of fire and light symbolism.

Light is not only sacred because it dispels the darkness in which lurk all the enemies of human life. It is also sacred because it is the vehicle of life. This is evidenced by the effect of sunlight upon plant, animal, and human life. Light is also the vehicle of color, the coloring matter of all earthly things being imparted from the sun. It is the vehicle of heat, and according to the wisdom of antiquity, it carries the sperm of all things from the sun. Through light also pass the impulses from the Grand Man. According to the Mysteries, God controls His universe by means of impulses of intelligence which He projects through streamers of visible or invisible light. This light serves the universe in a capacity somewhat similar to that in which the nervous system serves the body.

Pythagoras said; "The body of God is composed of the substance of light." Where light is, God is. Who worships light, worships God. Who serves light, serves God. What more fitting symbol has any man ever conceived of the ever-living, pulsating Divine Father than the living, pulsating radiating fire. Fire is the most sacred of all elements and the most ancient of all symbols. This being the case, the ancients were not without reason and philosophy when they accepted fire, or light, as their supreme symbol, and chose as the emblem of the universal light the central glory of the sun. In so doing they became not sun-worshippers but worshippers of God as He manifests Himself through the light of truth.

The fire philosophers worshipped three lights—the light of the sun, that of the earth, and that of the soul, this latter being the light in man which they believed would ultimately be reabsorbed into the Divine Light from which it was temporarily separated by the prison walls of man's lower nature. The Mysteries of all ages were dedicated to the reunion of the little light with the Great Light, its Father and Source. To the Gnostics Christ was the colorless divine Light which assumed the form of radiant splendor (Truth), that it might minister unto the needs of the little light struggling for expression in the soul of every human creature. This Divine Light entered into the light of Nature, and by strengthening the latter, assisted the vitalizing of all living things.

The light in man, the God in miniature, was saved—or more correctly, *released*—by a process called *regeneration*. The secret method used to effect this release without the long spiral path of evolutionary progress was the great and supreme secret of the Mysteries, revealed only to those who had proven themselves worthy to be entrusted with the power of life and death. These Mysteries are perpetuated today in Freemasonry.

The Masonic order is founded in the Secret Schools of the pagan antiquity, many of the symbols of which are preserved to this day in the various degrees of the Blue Lodge and the Scottish Rite. Concerning the origin of the name "Freemason," which is itself a key to the doctrines of the Order, Robert Hewitt Brown, 32°, writes: "Long before the building of the King Solomon's temple, masons were known as the '*Sons of Light.*' Masonry was practiced by the ancients under the name of *Lux* (light), or its equivalent, in the various languages of antiquity. * * * We are informed by several distinguished writers that it (the word *Masonry*) is a corruption of the Greek word *Mesouraneo,* which signifies 'I am in the midst of heaven,' alluding to the sun, which, 'being in the midst of heaven,' is the great source of light. Others derive it directly from the ancient Egyptian *Phre*, the sun, and *Mas*, a child; Phre-Massen—children of the sun, or Sons of Light."

The true secret of the regeneration of the fire in the human soul is revealed by the ritual of the Third Degree of the Blue Lodge under the allegory of the murder of Hiram Abiff. The name *Hiram* is, as has already been noted, closely related to the element of fire. His direct descent from Tubal-Cain, the first great worker of metals by means of fire, still further connects this cunning worker of metals with the immortal life flame in man. In his *Secret Societies of all Ages,* Charles W. Heckethorne gives an old Qabbalistic legend in connection with the relation of early Masonry to the worship of fire. According to this legend, Hiram Abiff was not a descendant of Adam or Jehovah, as were the Sons of Seth, but was born of a nobler race, for in his blood ran the fire of Samael, one of the Elohim. Further, there are two kinds of people in the world; those with aspiration and those without. Those without aspiration are the Sons of Seth, true children of the earth, who cling to their parent with tenacity, and the keyword of their nature is *Earthiness*.

There is another race who are Sons of Fire, for they are descendants of Samael, the regent of fire. These flame-born sons are ever fired with ambition and aspiration. They are the builders of cities, the raisers of monuments, the conquerors of worlds, the pioneers, the workers in metals, true sons of the eternal flame. Fiery and tempestuous are their souls and earth to them is a burden. Jehovah does not answer their prayers, for they are sons of another star. *Aspiration* is the keynote of their nature, and again and again they raise, phoenix-like, from the ashes of failure. Never will they rest, like the element of which they are a part; they are wanderers upon the face of the earth, with their eyes upon the flaming star from which they came.

This fundamental difference is plainly visible in daily life. Some are always contented; others never reach the goal. Some are the Sons of Water—the keepers of flocks; others are Sons of Fire—the builders of cities. One group is conservative, the other is progressive. One is the king, the other the priest. But within the nature of every living thing the Sons of Fire and the Sons of Water exist together. In the Scriptures the flame-born ones are called the Sons of God and the water-born are referred to as the Daughters of Men, for the flame-born Son is the divinity in man and the water-born is the humanity in man. These two brothers are deadly enemies, but in the Mysteries they are taught to cooperate one with the other and are symbolized in Freemasonry as the double-headed eagle of the 33°.

According to the ancient wisdom, a time will come when man has two complete spinal systems, both equally developed, and his life will be controlled by two powers working in unity one with the other. To express this the ancient alchemists symbolized attainment as a two-headed figure, one head male and the other female. The androgynous *Ishwara,* the planetary lord of the Brahmins, has the right half of his body male and the left half female, to symbolize that he is the archetype for the ultimate human race. Man then being positive and negative in one, will no longer reproduce himself as at the present time.

One of the Ancient Mysteries taught that the end of all things is like the beginning plus the experience of the cycle, and some day the human race will give birth to its new bodies out of its own nature, as certain primitive animals still do. Then, indeed, will man be his own father and his own mother, complete in himself. Initiation makes possible this process in man much earlier than the natural sequence of human evolution would permit it. Such is the true mystery of Melchizedek, King of Salem, the Priest-King (Priest, water; King, fire) who was his own father and his own mother, and in whose footsteps all initiates follow.

The highest of all occult orders which exists only in the inner world may be called the "Order of Melchizedek," although among certain nations it has other names. This Order is composed entirely of the graduates of the other Mystery Schools who have actually reached the point where they can give birth to their present selves out of their own natures, like the mysterious phoenix bird which, breaking open at death, permits a new bird to fly forth. The Phoenix was once regarded as an actual zoological rarity, but it is now known that it never existed other than a symbol of a high stage in the development of man. The phoenix built its nest out of flames, which is exceedingly significant.

The secret Order of Melchizedek can never appear in the physical world while humanity is constituted according to its present plan. It is the supreme Mystery School, and a few have reached the point where they have blended their divine and human natures so perfectly that they are symbolically two-headed. The heart and mind must be brought into perfect equilibrium before true thinking or true spirituality can be attained. The highest function of the mind is reason; the highest function of the heart is intuition, a sensing process not necessitating the normal working of the mind. Reason alone is heartless; feeling alone is mindless, but these two blended together temper justice with mercy and kindliness with strength.

The spirit is neither male nor female, but both—an androgynous entity. The perfect manifestation of the androgynous spirit must be through an androgynous body, which is self generating. But many millions of years must pass before the human race learns the lessons of polarity sufficiently well to assume this new nature with intelligence. In that day everything will be complete unto itself. Understanding will be mature and there will be a depth and broadness which cannot be manifest through either a male or a female organism alone. Such is the mystery of the Priest-King and such was the position which Jesus reached when he was called a priest forever after the Order of Melchizedek. All this is symbolized in the emblems of the 33° of Freemasonry.

When considered clairvoyantly, the body of man resembles a great bouquet of flowers, for all over the physical form are petal-like groups of emanating force rays of various shapes and colors. There is one of these mysterious centers in the palm of each hand and in the sole of each foot. Nearly all the vital organs have whirling or radiating vortices of light as spiritual bases. These spinning vibrating flowers are extremely important occult centers. Each of them is capable, under certain conditions, of assisting man to secure a broader function of consciousness.

It is possible to see with the palms of the hands and the soles of the feet. In fact, ultimately man will see with all parts of his body. A symbol of this ultimate condition was preserved in the Egyptian Mysteries by the figure of Osiris, who is often shown sitting upon a throne, his entire body composed of eyes. The Greek god Argus was also noted for his ability to see with different parts of his body. The Oriental Buddhas are often symbolized as having peculiar geometric patterns on the palms of the hands and the soles of the feet. The famous footprints of Buddha carved in stone have a miniature sun just in front of the heel of each foot. Some of the Japanese jiu-jitsu artists are acquainted with the secret science of these mysterious nerve centers, although the knowledge has been concealed from the majority of the Japanese wrestlers. There are charts in Japan which show the exact location of these sacred centers. The slightest pressure upon certain of them will paralyze the entire body, so great is their control over the rest of the nerve system.

The jiu-jitsu exponents are also taught how they can resuscitate a person who is absolutely dead by means of pressure brought to bear on certain points in the upper vertebrae of the spine. This is successful in nearly every case, often after all other methods have failed.

The sunbursts of varicolored lights in the body constitute the sacred lotus blossoms of India and Egypt and the roses of the Rosicrucians. They are also the immortal beads of the Bhagavad-Gita strung upon a single thread. It is through these centers that the nails of the crucifixion were driven. The crucifixion contains the secret of opening the flower centers in the hands, feet, side, and head. The three nails which accomplish this are preserved to Freemasonry as the three leading officers of a Lodge and the three murderers of Hiram Abiff.

The Mexican Indian Osiris, called Prince Coh, died from three wounds inflicted by his enemies, and his heart was found in an urn by Augustus Le Plongeon, who spent many years in investigating Central American antiquities.

The relationship between these sacred centers and the jewels in the breastplate of the High Priest of Israel must not be overlooked, for both symbols have a similar meaning.

The most sacred part of the human body is the brain and spinal system, revered from all antiquity and symbolized again and again in all the religions of the world. While other parts of the body are of great interest to the student, the mysterious working of the spinal fires by means of which liberation is finally attained is so tremendous that many years must be spent in understanding even the fundamental principles. The spine is the rod which budded, the Yggdrasil Tree, the flaming sword, the staff of comfort, the wand of the Magi.

PART III

THE SACRED FIRE IN THE SPINE AND BRAIN

Santee called the *medulla spinalis* (spinal cord) the central axis of the nervous system. In a person of average size the spinal cord is about eighteen inches in length, weighs approximately one ounce, and terminates opposite the first lumbar vertebra. The upper end of the spinal cord, passing upward through the *foramen magnum* (the large opening in the occipital bone of the skull) ends at the *medulla oblongata.* Running through the spinal cord is a tiny central channel referred to as the *sixth ventricle.* This is described as follows; "It is just visible to the naked eye, but it extends throughout the cord and expands above the fourth ventricle. In the *conus medullaris* it is also dilated, forming the *ventriculus terminalis* (Krausei)."

According to the Eastern system of occult culture, there are 49 sacred nerve centers in the human body, of which the seven most important and key centers are placed near the spine at various intervals. The total number, 49, is the square of 7, and is also the number of rounds and sub-rounds of a planetary chain. When seen clairvoyantly, all of these centers resemble flowers or electric sparks. Each one of the seven main plexuses has six of lesser importance surrounding it, thus forming six-pointed stars diagrammatically, although the centers are not arranged in starlike order in the body.

Concerning the continued recurrence of the sacred number

seven in connection with the organs and parts of the human body, H. P. Blavatsky writes; "Remember that physiology, imperfect as it is, shows septenary groups all over the exterior and interior of the body; the seven orifices, the seven 'organs' at the base of the brain, the seven plexuses the pharyngeal, the laryngeal, cavernous, cardiac, epigastric [some as solar plexus], prostatic, and sacral plexus, etc." These seven are the negative spinal plexuses of first importance, but disciples of the mysteries are warned not to attempt the development of these centers, because they are negative poles. All of the real plexuses which the true disciple of the highest knowledge should try to unfold are located within the skull, for the body is a negative pole of that spiritual body contained within the cranial cavity. As the body is controlled by the brain, the true adept works with the brain, avoiding the negative poles of the brain centers which are located along the spine. Proper development of the seven brain discs, or spiritual interpenetrating globes, results in the awakening of the spinal flowers by an indirect process. *Beware of the direct process by concentrating upon or directionalizing the internal breath towards the spinal centers.*

Madame Blavatsky might have added to her list of septenaries the seven sacred organs about the heart, the seven ductless glands of first importance, the seven methods by which the body is vitalized, the seven sacred breaths, the seven body systems (bones, nerves, arteries, muscles, etc), the seven layers of the auric egg, the seven major divisions of the embryo, the seven senses (five awakened and two latent), and the seven-year periods into which human life is divided. All of these are reminders of the fact that seven primitive and primary spirits have become incarnated in the composite structure of man and that the Elohim are actually within his own nature, where from their seven thrones they are molding him into a septenary creature. One of these Elohim, which corresponds to a color, a musical note, a planetary vibration, and a mystical dimension, is the key consciousness of every kingdom in Nature. The Elohim also take turns in controlling the life of the human being.

According to the ancient Brahmins, the Lord of the human race is keyed to the musical note fa, and His vibration runs through the minute tube in the spinal column. This tube is called the *sushumna.* The essence moving through the *sushumna* finally blossoms outward, forming a magnificent flower in the brain. This is called *sahasrara,* the thousand petaled lotus, in the midst of which is enthroned the divine eye of the gods. In India it is possible to secure inexpensive chromos showing a meditating Yogi with these flower centers along the spine connected together by the three *nagas,* or serpent gods, which represent the divisions of the spinal cord. The caduceus of Hermes shows the two serpents wound around the central staff, where they vibrate as the sharp and flat notes of the central stem.

The *naga* gods, or serpents, often symbolized with human heads (sometimes as cobras with seven heads) are favorite motifs in Oriental art. In an isolated part of the jungle in Indo-China stand the remains of the ancient city of Angkor, concerning the building of which nothing is known, although the natives maintain that its great structures were raised in a single night by the gods. These buildings contain hundreds of carvings of great serpents, most of them hooded cobras. In some cases the bodies, being of great length, are used as railings around walls and the sides of steps. In the Indian chromos the blossoms along the spine are often shown with varying numbers of petals. For example, the one at the base of the spine has but four petals; the next above six. Each of these petals is inscribed with a mysterious Sanskrit character representing a letter of the ancient alphabet. The petals are believed to indicate the number of nerves branching from the plexus or ganglion.

The lotus blossoms are often ornamented with the figures of the gods, for all of the deities of the Brahmin Pantheon are related to nerve centers in the human body, and the proclivities which they demonstrate mythologically are symbolic of activities within the nature of man. One Oriental painting shows three sunbursts, one covering the head, in the midst of which sits Brahma with four heads, his body a dark mysterious color. The second sunburst, which covers the heart, solar plexus, and upper abdominal region, shows Vishnu sitting in the blossom of the lotus on a couch formed of the coils of the serpent of cosmic motion, its seven-hooded head forming a canopy over the god. Over the generative system is a third sunburst, in the midst of which sits Shiva, his body a grayish white, and the Ganges river flowing out of the crown of his head. This painting was the work of an Indian mystic who spent many years on the figures, subtly concealing therein great truths. The Christian legends could be related to the human body by the same method as the Oriental, for the meanings concealed in the teachings of both schools are identical.

In Masonry the three sunbursts represent the gates of the Temple, at which Hiram is struck, there being no gate in the north because the sun never shines from the northern angle of the heavens. The north is the symbol of the physical because of its relation to ice—crystallized water—and to the body—crystallized spirit. In man the light shines toward the north but never from it, because the body has no light of its own; it shines with the reflected glory of the Divine life particles concealed within the physical substances. For this reason the moon is accepted as the symbol of man's physical nature. Hiram, or Chiram as he should more properly be called—inasmuch as his name consists of the letters which in Hebrew stand for fire, air, and water—represents the mysterious fiery airy water which must be raised through the three grand centers symbolized by the ladder with three rungs and the sunburst flowers mentioned in the description of the Indian painting. It must also pass upward by means of the ladder of seven rungs, the seven lotus blossoms first described. These blossoms need not be considered entirely from an Oriental angle. Christians could properly call them the stations of the cross, for they are sacred places where the redeeming fire stops for a moment on its way up Calvary to liberation.

The spinal column is a chain of thirty-three segments, divided into five groups: (1) the *cervical,* or neck, vertebrae, seven in number, (2) the *dorsal,* or *thoracic*, vertebrae, of which there are twelve (one for each rib), (3) the *lumbar* vertebrae, five in number, (4) the *sacrum* (five segments fused into one bone), and (5) the *coccyx* (four segments considered as one). The nine segments of the *sacrum* and *coccyx* are pierced by ten *foramina*, through which pass the roots of the Tree of Life. Nine is the sacred number of man and there is a great mystery concealed in the *sacrum* and *coccyx.* That part of the body from the kidneys downward was called by the early Qabbalists the *Land of Egypt*, into which the Children of Israel were taken during the captivity. Out of Egypt Moses (the illuminated mind, as his name signifies) led the tribes of Israel, the twelve faculties, by *raising* the brazen serpent in the wilderness upon the symbol of the Tau Cross. At the base of the spine there is a tiny nerve center concerning which nothing is known, but the occultist realizes that the symbolism of the second crucifixion, which is supposed to have taken place in Egypt, has reference to the crossing of certain nerves at the base of the spine. A friend visiting Mexico was good enough to count the rattles on the tails of stone images of Quetzalcoatl, or Kulkulcan as he is sometimes known. In nearly every case they were nine in number.

The cosmic hierarchy controlling the Constellation of Scorpio has control of the occult fires in the human body. To symbolize this they were given the name of the *serpent gods* and the priests initiated into their mystery wore the coiled serpent in the form of the *uraeus* upon their foreheads. These priests also often carried flexible staffs carved in the form of a snake and from six to ten feet long. The wood of which they were made was specially treated by a process now lost. At a certain part of the ceremonial, the priests bent the flexible staffs into circles, placing the tail of the carved snake into its mouth, accompanying the process with secret invocations. The transcendentalists of the Middle Ages did the same thing, but not with the full understanding possessed by the ancient priests. The Lords of Scorpio, being the great initiators, accepted none into the Mysteries save when the sun was in a certain degree of Taurus, symbolized by Apis, the Bull. When the Bull carried the sun between his horns, the neophytes were admitted. In geocentric astrology, this takes place when the sun is supposedly in the last decan of the Constellation of Scorpio. This is true not only in the ancient Egyptian rituals, but it is still true in the Mystery Schools. Candidates for the occult path of fire are to this day admitted only when the sun is geocentrically in Scorpio and heliocentrically in Taurus. The star group constituting the Constellation of the Scorpion closely resembles a spread eagle and is one of the reasons why that bird is sacred to Freemasonry, which is a fire cult.

Although the three tubes of the spinal cord are called in the ancient wisdom the *nagas*, or *whirling snakes*, and the serpent which cannot die till sun-down was accepted as their symbol, the scorpion has also been used as emblematic of the spinal fire. This scorpion was called Judas, the betrayer, for he is a backbiter, carrying his sting in the *sacrum* and *coccyx.* We are reminded of the legend of Parsifal, for the Castle of Klingsor, the evil magician, located at the foot of the mountain in the midst of a garden of illusion, is merely a symbol of this City of Babylon and the Land of Darkness, where all too often the Son of God is tempted to sacrifice his immortality. Here also is the scene which Goethe called "Walpurgis Night." It is here also that the false light is chained for a thousand years, as related by

Milton in *Paradise Lost.*

Concerning the descent of the spirit of fire down the spine into the place of darkness, Milton says: "Him the Almighty Power hurl'd headlong flaming from the ethereal sky, with hideous ruin and combustion, down to bottomless perdition, there to dwell in adamantine chains and penal fire!"

It is also from here that the hordes of scorpions arose, spreading blight and destruction to all the earth, as is related in the Book of Revelation. Here also is the rock Moriah, over the brow of which Hiram is buried. Here lurks Typhon, the Destroyer, and Satan, who was stoned. This is the dwelling place of the false light, to differentiate it from the true light which shines out through the regions of *schamayim* above. Between these two lies the length of the spinal cord, a rope connecting the Ark and the Anchor.

There is a legend among the Orientals to the effect that Kundalini, the goddess of the serpentine spinal fire, growing tired of heaven, decided to visit the new earth which was being formed in the sea of space. She therefore climbed down a rope ladder (the umbilical cord) from heaven and found an island (the fetus) in the Sea of Meru (the amniotic fluids) surrounded by the mountains of Eternity (the Chorion), all of which existed within the egg of Brahma (the womb of Matripadma). After exploring the island, Kundalini decided to return up the ladder once more, but as she was about to ascend, the ladder was cut from above (the umbilical cord severed at birth) and the island drifted off into space. In fear Kundalini ran and hid herself in a cave (the sacral plexus) where, according to certain of the Eastern teachings, she remains coiled like the cobra in the snake charmer's basket, from which she can be lured only by the three mysterious notes of the charmer's flute. When Kundalini begins to unwind she ascends as a serpentine stream of fire through the spinal canal and into the brain, where she stimulates the activity of the pituitary body.

The spine may be divided horizontally into three sections. The lowest section includes the *lumbar* vertebrae, together with the segments forming the *sacrum* and *coccyx*, and is surrounded by a brick-red haze of a lurid and angry color. This haze is oily in texture and causes the *sacrum* and *coccyx* to appear the color of dried blood. The color, however, is living rather than dead. Higher up towards the lumbar vertebrae the color is somewhat lighter and not so angry looking. It gradually turns to orange and through the section composed of the twelve *dorsal* vertebrae there is a golden glow radiating outward from a thread-like line of what appears to be a river of yellow fire. These colors extend somewhat outward along the nerves which branch off from the spine between the vertebrae. A little higher the yellow becomes flecked with green and through the *cervical* section the stream becomes faintly electric blue. Through the *ida* and *pingala*—two lateral tubes through the spinal cord, paralleling the central tube on either side—this stream of fire goes up and down incessantly. The farther up the fire goes, the thinner and less brilliant its hues, but the purer and more beautiful the colors, until finally they meet in a seething, molten mass in the *pons* of the *medulla oblongata*, where the fire begins almost immediately to permeate the third ventricle and agitate the pituitary body.

This tiny form is described by Santee as follows: "The *hypophysis* (pituitary body) is composed of two lobes bound together by connective tissue. A sheet of *dura mater* (diaphragma sellae) holds them in the hypophyseal fossa. The anterior lobe, the larger, is derived from the epithelium of the mouth cavity; and, in structure, resembles the thyroid gland. Its closed vesicles, limed with columnar epithelium (in part ciliated), contain a viscid jelly-like material (pituita), which suggested the old name for the body. The anterior lobe is hollowed out on its posterior-surface (kidney-shape) and receives the posterior lobe, the infundibulum, into the concavity. The hypophysis has an internal secretion which appears to stimulate the growth of connective tissues and to be essential to sexual development."

The pituitary body is the negative pole, yet it plays many roles in the development of the spiritual consciousness. In one sense of the word it is the initiator, for it "raises" the candidate (the pineal gland). Being of feminine polarity, the pituitary body lives up to its dignity by being the eternal temptress. In the Egyptian myths, Isis, who partakes of the nature of the pituitary body, conjures Ra, the Supreme Deity of the sun (who is here symbolic of the pineal gland) to disclose his sacred name, which he finally does. The physiologic process by means of which this is accomplished is worthy of detailed consideration.

The pituitary body begins to glow very faintly and little rippling rings of light pour out from around the gland and gradually fade out a short distance from the pituitary body. As occult growth continues, according to the proper understanding of the law of occultism, the emanating rings around the gland grow stronger. They are not equally distributed around the pituitary body. The circles are elongated on the side facing the third ventricle and reach out in graceful parabolas towards the pineal gland. Gradually, as the stream becomes more powerful, they approach ever closer to the slumbering eye of Shiva, tinting the form of the pineal gland with golden-orange light and gently coaxing it into animation. Under the benign warmth and radiance of the pituitary fire, the divine egg thrills and moves and the magnificent mystery of occult unfoldment takes place.

The pineal gland is thus described by Santee: "Pineal body (*corpus pineale*) is a cone-shaped body, 6mm. (0.25 in.) high and 4 mm. (0.17 in.) in diameter, joined to the roof of the third ventricle by a flattened stalk, the *habenula.* It is also called the *epiphysis.* The pineal body is situated in the floor of the transverse fissure of the cerebrum, directly below the splenium of the corpus callosum, and rests between the superior colliculi of the quadrigeminal bodies on the posterior surface of the midbrain. It is closely invested by pia mater. The habenula splits into a dorsal and ventral lamina, which are separated by the pineal recess. The ventral lamina fuses with the posterior commissure; but the dorsal stretches forward over the commissure in continuity with the roof epithelium. The border of the dorsal lamina is thickened along the line of attachment to the thalamus and forms the *stria medullaris thalami* (pineal stria). The thickening is due to the presence of bundle of fibres from the columna of the fornix and the intermediate stria of the olfactory tract. Between the medullary striae at the posterior end there is a transverse band, the *commissura habenularum*, through which the fibres of the striae partially decussate to the nucleus habenulate in the thalamus. The interior of the pineal body is made up of closed follicles surrounded by ingrowths of connective tissue. The follicles are filled with epithelial cells mixed with calcareous matter, the brain-sand (*acervulus cerebri*). Calcareous deposits are found also on the pineal stalk and along the chorioid plexuses.

"The function of the pineal body is unknown. Descartes facetiously suggests that it is the abode of the spirit (the sand) of man. In reptiles there are two pineal bodies, an anterior and a posterior, of which the posterior remains undeveloped, but the anterior forms a rudimentary, cyclopean eye. In the Hatteria, a New Zealand lizard, it projects through the parietal foramen and presents an imperfect lens and retina and, in its long stalk, nerve fibers. The human pineal body is probably homologous with the posterior pineal body of reptiles."

The pineal gland is a link between the consciousness of man and the invisible worlds of Nature. Whenever the arc of the pituitary body contacts this gland there are flashes of temporary clairvoyance, but the process of making these two work together consistently is one requiring not only years but lives of consecration and special physiological and biological training. This third eye is the Cyclopean eye of the ancients, for it was an organ of conscious vision long before the physical eyes were formed, although vision was a sense of cognition rather than sight in those ancient days.

As man's contact with the physical world grew more complete he lost his inner understanding together with the conscious connection with the creative hierarchies. In order to regain this connection, it is necessary for him to rise above the limitations of the physical world. He must not, however, sever his connection with humanity by becoming a recluse or an impractical dreamer. The occultist must not *walk* out of anything; he must *work* out of everything. In the pineal gland there is a tiny grit, or sand, concerning which modern science knows practically nothing. Investigations have shown that this grit is absent in idiots and others lacking properly organized mentality. Occultists know that this grit is the key to the spiritual consciousness of man. It serves as a connecting link between consciousness and form.

The foregoing will give brief understanding of part of the mystery of man's occult anatomy. Those with discerning eyes will see in the spinal canal leading upward into ventricles of the brain—through certain doors concerning which science is ignorant—the passageways and chambers of the ancient Mysteries. They will realize that the spinal spirit fire is the candidate who is being initiated. In the triangular room of the third ventricle the Master Mason's Degree is given. Here the candidate is buried in the coffin; and here, after three days, he rises from the dead.

The lower degrees of the ancient Mysteries led through tortuous passageways where howling and crying rent the air and the flames of the inferno flickered through the darkness. The neophyte seeking for light was first led through the underworld, where he fought strange beasts and heard the wailing of lost souls. All this was emblematic of man's own lower nature, through which his spiritual ideals must rise to illumination and truth. The higher degrees of the Mysteries took place in beautiful domed buildings where white-robed priests chanted and sang, and lights from invisible lamps shone on golden jewels. The Greater Mysteries represented the felicity of the soul surrounded by light and truth. They symbolized that man had "raised" himself from the darkness of ignorance into the light of philosophy. Plato said that the body is the sarcophagus of the soul, for he realized that within the form was an immortal principle which could free itself from mortal sheath only by death or by initiation. By the ancients these two were considered almost synonymous. For that reason the dying Socrates ordered his disciples to make an offering at the time of his death similar to the one which candidates made when about to be initiated into the Eleusinian Mysteries.

The mystery of the All-Seeing Eye was sometimes symbolized by the peacock, because this bird had eyes in all of its feathers. In honor of the sacred eye in the crown of the head, the monks of all nations shave their hair over the place which it is supposed to look out. Small children who have but recently completed their embryonic recapitulation of humanity's early struggle for life have and unduly sensitive area about the crown of the head. The skull does not close there immediately. In some cases it never closes, although usually the sutures unite between the second and fifth years. The extreme sensitiveness over the area of the third eye is accompanied by a certain clairvoyance. The small child is still living largely in the invisible worlds. While its physical organism is unresponsive, it is conscious and active (at least to limited degree) in those worlds with which it is connected by the open gateway of the pineal gland. Gradually certain manifestations of his higher consciousness enter into its physical organism and crystallize into the fine grit found in that gland. There is no grit in the pineal gland until consciousness enters.

Not only are the two glands in the head tremendously important, but the whole glandular system, especially the ductless glandular system, exercises tremendous sway over the human system. The white blood corpuscles are not actually manufactured in either the pancreas or the spleen, but are really formed by activity of the etheric double, which is connected to the physical form through the spleen. A continuous stream of partly etheric white blood corpuscles pours from the invisible world into the visible organism through the gateway of the spleen. The same is true of the liver, for the red blood corpuscles are to a certain degree a crystallization of astral forces, for the liver is the portal leading into the astral body.

The seven major ductless glands are under the control of the seven planets, and each one of them is actually a sevenfold body in the same way that all the vital organs are sevenfold. The heart is divided into seven complete organs by a certain process of occult anatomy, as is also the brain. The fact that the human brain closely resembles in certain details—especially the organs grouped about the base of it—an androgynous human embryo, is sufficient to cause further investigation. Sometime physicians will realize that the knowledge of the organs and functions of the human body is the most important and complete method of understanding the religions of all the world, for all religions—even the most primitive—are based on the functions of the human form. It was not without reason that the ancient priests placed over the temple gate the immortal motto:

"MAN KNOW THYSELF"

UNIVERSITY OF PHILOSOPHICAL RESEARCH
2001